Jacques Cousteau

and the Undersea World

Explorers of New Worlds

Daniel Boone and the Exploration of the Frontier
Jacques Cartier and the Exploration of Canada
Christopher Columbus and the Discovery of the New World
Francisco Coronado and the Exploration of the American Southwest
Hernando Cortés and the Conquest of Mexico
Jacques Cousteau and the Undersea World
Sir Francis Drake and the Foundation of a World Empire
Vasco da Gama and the Portuguese Explorers
Sir Edmund Hillary: Modern-Day Explorer
La Salle and the Exploration of the Mississippi
Lewis and Clark: Explorers of the Louisiana Purchase
Ferdinand Magellan and the First Voyage Around the World
Francisco Pizarro and the Conquest of the Inca
Marco Polo and the Wonders of the East
Juan Ponce de León and the Search for the Fountain of Youth
Reaching for the Moon: The Apollo Astronauts
Theodore Roosevelt and the Exploration of the Amazon Basin
Sacagawea: Guide for the Lewis and Clark Expedition
Hernando de Soto and the Exploration of Florida
The Viking Explorers

Explorers of New Worlds

Jacques Cousteau

and the Undersea World

Roger King

Chelsea House Publishers
Philadelphia

Prepared for Chelsea House Publishers by:
OTTN Publishing, Stockton, N.J.

CHELSEA HOUSE PUBLISHERS
Production Manager: Pamela Loos
Art Director: Sara Davis
Director of Photography: Judy L. Hasday
Managing Editor: James D. Gallagher
Senior Production Editor: J. Christopher Higgins
Series Designer: Keith Trego
Cover Design: Forman Group

http://www.chelseahouse.com

First Printing
1 3 5 7 9 8 6 4 2

Library of Congress Cataloging-in-Publication Data

King, Roger.
Jacques Cousteau and the undersea world / Roger King.
p. cm. – (Explorers of new worlds)
Includes bibliographical references and index.
ISBN 0-7910-5956-1 (hc) – ISBN 0-7910-6166-3 (pbk.)
1. Cousteau, Jacques Yves–Juvenile literature. 2. Oceanographers–France–Biography–Juvenile literature. [1. Cousteau, Jacques Yves. 2. Oceanographers.] I. Title. II. Series.

GC30.C68 K54 2001
551.46'0092–dc21
[B] 00-034592

Contents

The Man and the Sea

The French marine explorer Jacques-Yves Cousteau stands in front of a giant model of a blue whale on the Champs-Elysées in Paris. The whale was being used to promote Cousteau's undersea museum.

I

Since people first set sail on the open seas, they have wondered what lies beyond sight of land. Indeed, it was often the prospect of discovering new trade routes, or strange lands that held fabulous riches, that lured explorers to venture from the safety of their home ports.

Although these early mariners discovered entire new continents and proved facts we now take for granted (such as that the world really *is* round), in many ways their sailing ships merely skimmed the surface of the earth's vast seas, which cover three-fourths of our planet. Not until well into the 20th century would curious explorers take advantage of technology that allowed them to study what lay beneath the surface of our watery world as well.

No one is more widely credited with leading this exploration of the undersea world than Jacques-Yves Cousteau. This Frenchman pioneered skin diving after helping to perfect the ***aqualung***, which a person could use to breathe underwater.

Before the invention of diving gear, people could stay underwater only as long as they could hold their breath. Some natives of islands in the Pacific were trained to dive for pearls from childhood. They could stay under for a very long time–Cousteau himself documented a 60-year-old man who dove 130 feet deep in two and a half minutes–but these divers were the exception rather than the rule.

The first diving equipment relied on airtight diving suits, lead boots, heavy steel helmets, and umbilical-cord-like hoses that provided air. In these cumbersome outfits, the divers resembled astronauts in space–tethered to a mother ship for the very air they needed to sustain life.

The development of the aqualung allowed divers to explore the ocean floor freely.

Cousteau became convinced that skin divers would need a portable breathing device they could carry on their backs for freedom of movement. The equipment would have to contain enough air to sustain divers for more than a few minutes if they were ever to thoroughly explore the marine world. Cousteau wanted to allow divers to range through the water as freely as a bird flies through sky. To this end, he and others perfected diving gear, including masks, fins, diving suits, and the aqualung.

Although helping to develop the aqualung would be one of Jacques Cousteau's most notable achievements, the relentless undersea explorer also accomplished a feat even more noteworthy. Beginning in 1968, his movies of ocean life brought never-before-seen images of whales, sharks, dolphins, and other fascinating sea creatures into the living rooms of millions of television viewers for eight years.

Previously, the closest most Americans ever got to such creatures was reading about them, seeing them in an aquarium, or perhaps visiting the beach during the summer. However, thanks to Cousteau's development of underwater cameras and filming techniques, generations of landlubbers could finally imagine how exhilarating it could be to touch a humpback whale's skin as it glided serenely by, or stare through shark-cage bars as a great white bore down to investigate out of the deep blue depths.

Cousteau's landmark work in marine exploration involved 80 expeditions, some one million miles traveling around the globe, and the production of 65 films, including two that earned him Academy Awards: *The Silent World* and *World Without Sun.* In addition, he wrote or cowrote

Another term for diving gear like the aqualung is *scuba*, which stands for *S*elf-*C*ontained *U*nderwater *B*reathing *A*pparatus.

more than 50 books published in over a dozen languages. But perhaps his most lasting legacy was founding the Cousteau Society. This international ***nonprofit organization*** is dedicated to the study and preservation of our Water Planet, as Cousteau liked to call Earth.

Indeed, during the last decades of his life, Cousteau made it his mission to warn humankind of the damage pollution, overfishing, and thoughtless coastal development were wreaking on the marine world. The undersea explorer, who died in 1997 at age 87, was determined to make people aware that our practices on land–farming, industry, and housing–have a direct effect on the health of the watery world. Unlike the earth's landmasses, its oceans and seas connect nation to nation and people to people–whether they are aware of it or not.

One of his biggest fears was that current generations were not taking steps to preserve the world's natural resources for generations to come, an attitude he said had to change if our species was to survive. "The future of civilization depends on the water," he once said. "I beg you all to consider this. You all now have the duty, you have the time to convince people [to conserve natural resources]."

Risky Business

Jacques Cousteau's diving partners help him with his equipment before a dive. Cousteau and Emile Gagnan developed the aqualung in 1943, during the Second World War.

2

For a man whose name would become synonymous with water, Cousteau's first choice was to see the world by air. In fact, he was close to graduating from flight school when he was seriously injured in a car accident. He broke 12 bones in all, including 5 fractures in his left arm. Doctors fully intended to cut off his paralyzed right arm until a raging infection in it subsided.

Cousteau had been born in 1910 in St. Andre de

Cubzac, France. When he was 20 years old, he graduated from the French naval academy at Brest and became an officer in his country's navy.

While recovering from his accident (Cousteau ended up a gunner, not a pilot), he was stationed in Toulon, a scenic French Mediterranean port. There he met three people who would greatly influence his life. One was a fellow navy officer named Philippe Talliez, who encouraged Cousteau to swim to hasten his recuperation. Talliez helped his friend to see underwater by giving him a pair of diving goggles. In Toulon, Cousteau also became friends with a renowned spear fisherman named Frederic "Didi" Dumas, who would become like a brother. The third person he met was Simone Melchoir, a French admiral's daughter he would marry in 1937.

Jacques Cousteau's undersea odyssey did not really begin until the late 1930s and early 1940s. Like much of Europe, France had been invaded by German armies during World War II and found itself an occupied country. Nonetheless, Cousteau used the time to begin experimenting with underwater breathing devices, especially after struggling with fellow naval divers for several hours to free a steel cable wrapped around a torpedo boat's pro-

Cousteau met his wife, Simone, while recovering from an accident. They were married in 1937.

pellers. Holding one's breath while trying to work under the surface was just too exhausting for even trained frogmen to do, he discovered.

Cousteau tested several different underwater breathing apparatuses. He also tried different air mixtures, including pure oxygen. On a few occasions, he and his companions nearly died during these experimental dives, either because the air they were breathing was dangerous beyond certain depths (as is the case with pure oxygen) or their equipment failed. Indeed, each time they dove underwater was a new and potentially dangerous experience. Even scientists at the time didn't completely understand how the human body is affected by ***water pressure*** at certain depths.

Although evolution theory holds that all life on earth initially sprang from the sea, the average person is much like a "fish out of water" when he or she dives below the surface without special equipment. For starters, unlike fish, humans cannot "breathe" the dissolved oxygen in water; therefore, we need to bring an air supply down with us. Also, because water is 800 times more dense than air, as humans descend into the sea they can feel the water pressure getting stronger.

Water pressure actually compresses, or squeezes, certain objects underwater, including hollow spaces in the human body, such as lungs (which are like balloons) and the ear canal. It's critical, therefore, for divers to equalize the pressure on these hollow spaces. Skin divers do this in their ears by holding their nose and swallowing as they go deeper, allowing their ears to "pop" much the same as they would when rising through the atmosphere in an airplane.

The pressure that water places on an object 33 feet below the surface is equal to twice the atmospheric pressure at sea level. Thus, 33 feet below the water's surface a diver feels twice as much pressure as he or she would standing on the beach. This water pressure increases by one atmospere every 33 feet. It is three times as great at 66 feet, four times as great at 99 feet, and so on.

Diving underwater can also pose a problem because pressure changes the way that gases react in our bodies. Nitrogen, which makes up almost 80 percent of the air we breathe, is not exhaled from a diver's body underwater as readily as on land. In fact, water pressure can actually make nitrogen gas dissolve into the bloodstream instead of being expelled along with carbon dioxide. If a diver returns too quickly to the surface, where the water pressure is less strong, the nitrogen in the bloodstream will expand. These bubbles of nitrogen gas in the blood cause a painful condition known as the

The diver's movement and range were constricted in the original diving suits, like the one shown here. The development of the aqualung allowed people to move about more freely under the water.

bends. This name came from the twisted positions of divers who had come to the surface too quickly.

To avoid the bends, undersea explorers eventually learned that they had to stop periodically at certain depths when returning to the surface. There, they would wait to ***decompress***, or allow the compressed gases in their bodies to gradually work their way out of muscles, joints, and blood vessels. Divers who failed to decompress and were stricken with the bends when they reached the surface had to be put into pressurized decompression chambers immediately, or they could suffer irreversible injuries and even die.

Despite all the risks of scuba diving, Cousteau and his friends continued to test the limits of the new aqualung. Frederic Dumas was an extraordinary risk-taker. He became the first man ever to scuba dive to a depth of 210 feet and return unharmed to the surface. The divers also began to test photography equipment underwater. They wanted to show people on land all the marvels they were seeing firsthand. Like pioneers in other fields, they had to take existing technology and modify it to come up with satisfactory results. Even though it was the early 1940s, and a war was raging across Europe, Cousteau and his associates managed to make crude underwater movies.

It was one of these films that convinced Cousteau's naval superiors to permit him to continue diving experiments after the Nazis were driven out of France. His new outfit was called Groupe de Recherches, Sous-Marines (Undersea Research Group). The men of the group undertook some strange research–including testing the effects of underwater explosions on themselves! They also were ordered to search for and retrieve underwater explosives, called ***mines***, that had been laid during the war by Germany. In addition, they filmed

French submarines laying mines on the ocean floor. Once, Cousteau's good friend Frederic Dumas rode a descending sub to the bottom and signaled its captain to release his mines by pounding on the hull with a hammer. The group's most somber duty, however, was probably the day its divers had to locate the wreckage of a navy plane 120 feet deep and retrieve the bodies of its dead crewmen, who were lying peacefully on the ocean floor.

Cousteau and his compatriots nearly lost their lives during two dives that weren't even in the ocean! Members of the Undersea Research Group made a pair of experimental dives into a famous freshwater spring that mysteriously bubbled to life once a year and flooded a nearby river. The French divers explored two deep vertical caves under the spring's surface. Both times, Cousteau and his partners fell victim to what they felt was "rapture of the deep." This is a condition called ***nitrogen narcosis***, in which the nitrogen gas in the diver's body causes him or her to become disoriented and confused.

Under the dreamy effects of nitrogen narcosis, divers have been known to experience all sorts of strange hallucinations. Some have even pulled out their air hoses to give a passing fish a breath, or cut their safety lines to the surface. But it wasn't rapture of the deep that had nearly drowned Cousteau and his fellow divers. They had not descended to the depth that normally triggers the dangerous effects of nitrogen narcosis. Instead, Cousteau discovered that the air compressor used to fill his men's tanks had been sucking in its own exhaust smoke. As a result, small amounts of carbon monoxide were being pumped into the aqualungs. Carbon monoxide gas can be deadly if it is breathed on land, and its effects are magnified under pressure.

Cousteau and the other members of the group simply modified the air compressor to keep the poisonous gas out of the tanks. With that problem solved, they returned to diving.

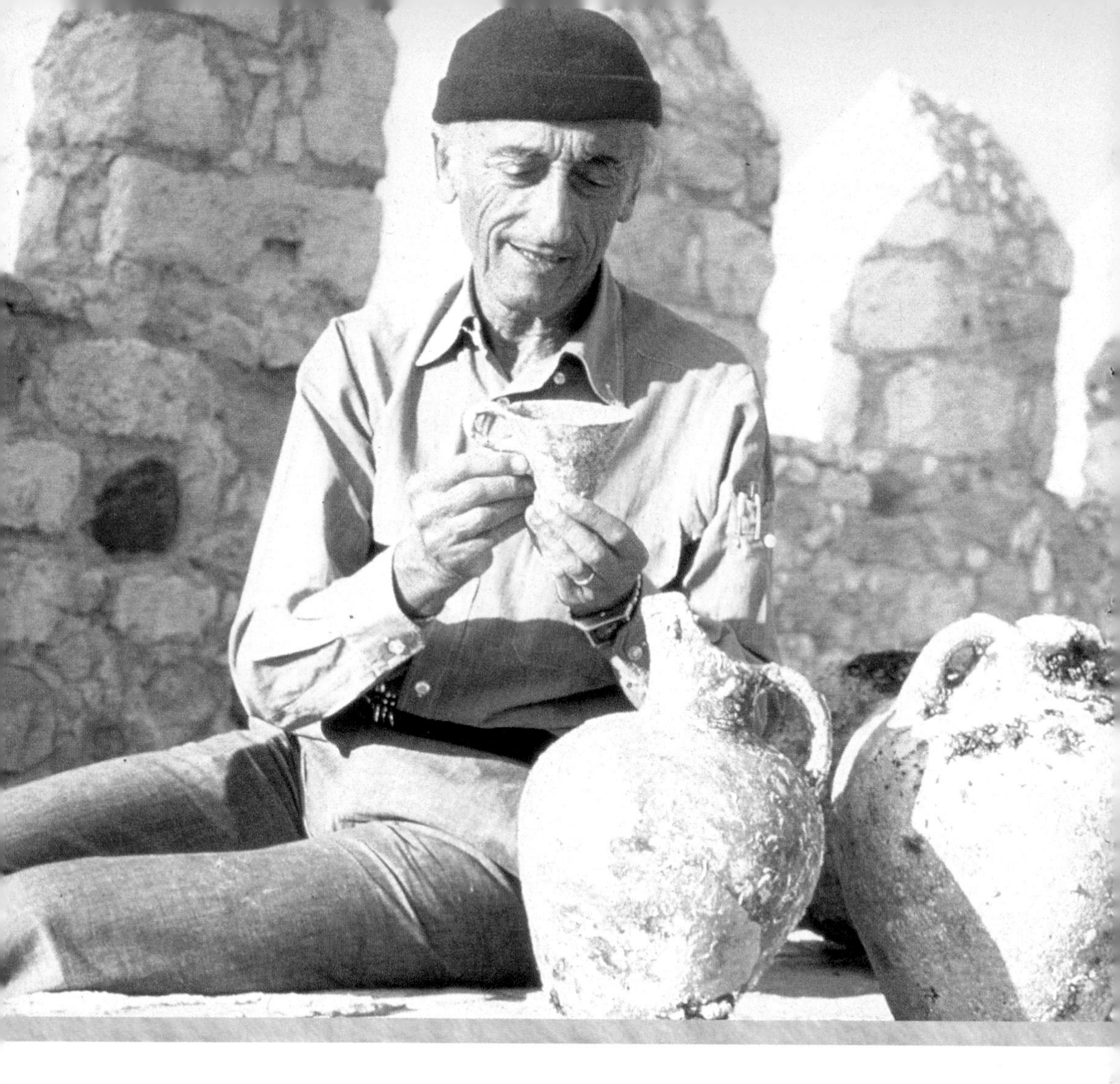

Wonders of the Deep

Jacques Cousteau takes a careful look at some ancient Greek pottery recovered from a long-forgotten shipwreck in the Mediterranean.

3

Cousteau's dives on behalf of France naturally took him into the Mediterranean Sea. This historic body of water has been traveled by countless sailors. It is 2,400 miles long, nearly three miles deep in some places, and surrounded by Africa, Asia, and Europe. The Mediterranean is a treasure trove of sunken ships and, in some cases, submerged civilizations. Some had remained hidden for thousands of years. To Jacques Cousteau, the

Mediterranean was a natural history museum. The only price of admission was staying underwater long enough to probe its mysteries.

In 1948, the Undersea Research Group received permission to find the ***argosy*** (a large merchant ship) of Mahdia. This ship had supposedly been used by a Roman emperor to carry home items from Greece. Among them was an entire Greek temple or building. It had been taken apart for the journey so that it could be rebuilt in Rome. Instead, some sudden storm or twist of fate had caused the ship to sink in over 100 feet of water.

In the early 1900s, divers using steel helmets and heavy suits had discovered the wreckage of the argosy. They had managed to bring up enough art objects to fill five rooms of a museum.

At first, Cousteau could not find the sunken ruins. Luckily, he at least knew the general area and depth of the ship's remains. It still took several days of tedious searching until a diver spotted a Greek column on the ocean floor. The wreck had been found once again.

The argosy of Mahdia rested 130 feet below the surface of the Mediterranean Sea. This was fairly deep, and the ***salvage operation*** would involve some risks. The Undersea Research Group divers could only stay underwater for 15 minutes at a time. Otherwise, they might succumb to the bends. To make sure divers followed the strict schedule, a rifleman on board the salvage vessel fired shots into the water to remind them when it was time to resurface.

There was just enough natural light reaching the wreck site for Cousteau to film the men at work raising marble columns and other large items from the argosy. They also found the remains of iron and bronze nails, as well as the perfectly preserved cedar ribs of the ship. Amazingly, the ship's coat of varnish was still intact after nearly 2,000 years.

In addition to diving on sunken wrecks, Cousteau and his team continued to test the limits of deep diving. They were anxious to learn exactly how and when divers were stricken by the mysterious rapture of the deep.

Before one dive, a 300-foot rope was dropped to the sea floor. Cousteau and his fellow divers planned to follow the rope as far down as they dared. They would write how they felt on white boards attached to the rope at different depths. They were also to jot down a message at the farthest depth they could safely reach.

Cousteau had reached the 200-foot mark when he was suddenly struck by nitrogen narcosis. In his delirious state, he debated exactly what color the "room" was around him, even though it had no floor or ceiling. He also described the nightmarish thoughts that raced through his mind: "I was ill in bed, terrorized with the realization that everything in the world was thick. My fingers were sausages. My tongue was a tennis ball. My lips swelled grotesquely on the mouth grip. The air was syrup."

He even imagined that he was looking at himself in a mirror, and the face he saw ordered him to go even deeper. Down Cousteau went, now even less in control of his thoughts and actions, all the way to the last white board, which was dangling at 297 feet. Summoning up all the strength and resolve he could muster in water pressure that was 10 times the pressure at the surface, he scribbled his name on the board with his indelible pencil.

Now free to return to the light of day, he dropped the 10-pound weight that had helped him descend and headed for the surface, stopping only to decompress. Cousteau thus became the first man to free-dive to nearly 300 feet. "It was impossible not to think of flying to heaven," he later wrote.

Several fellow divers managed the same feat, with varying degrees of aftereffects. One felt dizzy for an hour after resurfacing, and another had a 48-hour headache for his efforts. However, on a later deep-diving expedition, a group member named Maurice Fargues drowned after he pulled out his own mouthpiece while feeling the effects of rapture of the deep. His fellow divers discovered that their comrade had managed to scrawl an illegible message on a board 396 feet down, the deepest any scuba diver had ever gone.

As if scuba diving weren't exciting enough, early undersea explorers felt compelled to probe even deeper into the ocean depths. And in the late 1940s, the only way to do that was in a diving machine called a ***bathyscaphe*** (Greek for "depth craft"). Cousteau was invited to join bathyscaphe designer Auguste Piccard for a deep-sea dive off the coast of West Africa. Their planned depth: 13,000 feet, or nearly two and a half miles, the average depth of the world's oceans.

Piccard's bathyscaphe had an observation ball constructed of thick steel walls with two six-inch-thick Lucite windows. It had bright spotlights to illuminate the black depths. The craft was also fitted with powerful harpoon guns in case a giant squid decided to sample it for lunch. And it contained a unique air purification system that could support two men for up to a day underwater. The bathyscaphe also had one unnerving feature: it carried steel tanks filled with 10,000 gallons of special gasoline. Because the gasoline was less dense than water, it was used to bring the bathyscaphe back to the surface once its weights were jettisoned on the ocean floor. If a spark had triggered an explosion, the vessel could have been the first oceanic space launch. However, the bathyscaphe survived its initial test dives and eventually carried men to the deepest ocean reaches.

In 1962, Auguste Piccard's son Jacques and U.S. Navy lieutenant Don Walsh descended seven miles in the bathyscaphe *Trieste I.* They landed on the deepest point in the ocean floor, the Marianas Trench off the coast of Guam, then returned to the surface.

As he continued to break ground in deep-sea diving, Cousteau dispelled myths about "sea monsters" such as octopuses, moray eels, and sharks. Even Cousteau and partner Frederic Dumas were initially wary of octopuses, until they actually observed them in person, which meant picking them

Auguste Piccard's bathyscape is lowered into the ocean off the coast of Africa from the S.S. Scaldis *in 1948.*

up (the creatures seldom put their tentacle suckers on human skin) and even dancing with them. Usually, the creature's first instinct was to try to escape.

Moray eels had likewise earned a bad rap, as much for their appearance as for their tendency to snap their toothy jaws at anything after being hauled aboard a fisherman's boat. But like the octopus, the moray's reputation as a fierce guardian of sunken ships and dark caves was overblown. Cousteau's divers learned that an eel would bite an unwary diver probing undersea nooks and crannies, but they also realized the fierce-looking creatures were merely protecting their homes.

Barracudas were another fish that, while sporting dagger-like teeth, never threatened a diver on one of Cousteau's expeditions. Truth be told, Cousteau told interested listeners, the divers feared sea urchins (a common shellfish covered with long, sharp spines) far more than barracudas. The urchin's spines could break off in a diver's arm or leg, causing painful wounds and infections.

Equally troublesome for divers were fire coral and sea poison ivy, which caused painful allergic reactions on the skin of careless divers. These wounds could last for days.

Sharks were another matter altogether in Cousteau's experience. He initially considered them a threat mainly when a diver entered and left the water. Early in his diving career, therefore, he designed a rugged shark cage. In theory, he and his men could be lowered safely to the sea bottom in the cage, swim out to explore, and then return to the cage for a safe ride to the surface. However, the test dive in the cage did not go so well. Because the divers were not sinking as fast as the cage, they

Sharks were a frightening sight to divers. Cousteau and his men soon learned that the unpredictable creatures were especially dangerous when divers were entering and leaving the water.

banged their heads and tanks on its ceiling. He never used that shark cage again.

Just the same, Cousteau realized sharks could be unpredictable–particularly the known man-eaters such as the great white. But his confidence grew the day he and Dumas were startled to see a 25-foot great white swim toward them and then do an about-face the second it spied them. Maybe sharks were pussycats after all, he thought. But although he and his men had never been attacked in their early diving careers, they had been circled enough times by aggressive sharks to realize that one could never really tell what a shark would do.

In the 1950s Jacques Cousteau, shown here on his research ship Calypso, *began to gain international recognition for his undersea discoveries.*

Ahoy Calypso

4

In 1950, Jacques Cousteau was still working for the French government's Undersea Research Group. He was eager to conduct marine research on his own. However, to do this he needed a ship and special equipment. All this would be very expensive. Fortunately, a wealthy Englishman named Loel Guinness liked Cousteau's work. He agreed to ***finance*** the explorer's dream.

When Cousteau went looking for a ship, he found a World War II ***minesweeper*** that had been converted into a passenger ferry after the war. The ship had been renamed *Calypso*, in honor of a mythological Greek sea nymph.

Calypso *had been a minesweeping ship before being refitted as a research vessel. It was equipped with special cameras, navigation equipment, a crane, and even a small helicopter.*

Calypso was exactly what Cousteau was looking for, even though it would have to be totally refitted as a research ship. A crane was added to lift heavy loads aboard, space was converted into crew quarters, and the latest electronic ***navigation*** equipment was installed. In addition, many new features were added to the ship. One of these was an underwater viewing room with eight portholes. Cameras could film action under the surface from this room at the front of the ship.

Cousteau took leave from the French navy and formed a nonprofit group called the Campagnes Oceanographiques Francaises (French Oceanographic Expedition). With his wife and two young sons, Philippe and Jean-Michel, Cousteau cruised the Mediterranean Sea on *Calypso*'s maiden voyage in November 1951. At one point he and his crew dove on the coral reefs near the coast of Saudi Arabia, where they took extensive photographs of marine life that few people had ever seen.

In 1952, Cousteau explored a sunken wreck off the coast of Marseilles, France. It turned out to be a Roman ship that had sunk around 300 B.C. He would end up diving for two years on the ruins. He also began writing a book about his undersea experiences. He called the book *The Silent World.* First published in the United States in 1953, the book was an instant hit thanks to the author's riveting stories and color photographs of marine life. It would ultimately be published in 22 different languages and distributed around the world.

Life as an independent oceanographer was certainly exciting. However, Cousteau was constantly searching for people or organizations to help pay for his expeditions. In 1954 he made a deal with the French government. France would pay two-thirds of the cost to operate *Calypso.* In return, Cousteau's ship would carry French scientists on his expeditions for nine months of the year. France had its first oceanographic vessel, and Cousteau had his first long-term benefactor.

Thanks to his newfound financial health, the captain of *Calypso* could set out on a long journey. The ship traveled over 13,000 miles and visited the Red Sea, the Indian Ocean, and the coast of Madagascar. Cousteau and Frederic Dumas dove hundreds of times and filmed their underwater adventures. They ultimately produced a ***documentary film*** with the same title as his first book, *The Silent World.* The movie would earn several awards, including an Oscar in 1957.

This success thrust Jacques-Yves Cousteau even more into the international spotlight. He was asked to head the Oceanographic Institute of Monaco (Monaco is a tiny ***principality*** on the southeast coast of France). His success also convinced the 47-year-old navy officer that it was time to truly be independent. After nearly three decades of service, he retired from the French navy.

Cousteau continued to be as curious about what existed in the deepest ocean depths as astronauts were about the worlds beyond earth. Determined to explore underwater regions where no one had ventured, he worked to perfect small diving submarines. He called these vehicles "saucers," and hoped they would allow him to explore depths that not even the sun's rays could reach. After several less-than-successful experiments with remote-controlled minisubs, Cousteau spent several years devising a manned sub that would be more maneuverable than the early bathyscaphes designed by Auguste Piccard.

His first prototype was called *Hull No. 1*, a turtle-shaped sub fitted with a hatch, two windows, and exterior search lights to allow its team to brighten the ocean depths. It also sported a mechanical arm to retrieve objects from the ocean floor, and cameras to record the sights. The diving saucer's most unique engineering feature, however, was its jet nozzles, which replaced propellers and allowed the craft to maneuver nearly as well as a scuba diver.

Hull No. 1 easily survived an unmanned test dive to 2,000 feet. Unfortunately, it failed to survive being raised to the surface by *Calypso.* The hauling cable snapped and the expensive saucer sank in over 3,000 feet of water–too deep to be recovered.

Despite this disappointment, Cousteau set out to build a second diving craft, which he named *DS-2* (Diving Saucer 2). However, battery problems plagued the early test dives with the sub, causing numerous cabin fires. Ultimately, though, Cousteau and his team perfected *DS-2*, making over 1,000 dives with it to depths they had never before explored.

Jacques Cousteau had always fantasized about humans living beneath the sea. He feared overpopulation and pollution would one day force humankind to use the undersea world to live and grow food. So in 1962, the inventive marine explorer introduced the world to Continental Shelf Station number one (Conshelf I), an 8-foot-high by 17-foot-wide ***habitat*** for two people, submerged 35 feet near the coast of Marseilles. Staying inside the underwater "house"–complete with piped-in hot water–were divers Albert Falco and Claude Wesly, who spent a week underwater.

A crane moves the above-surface section of Jacques Cousteau's undersea habitat ConShelf II into position. Cousteau's three experimental habitats proved that people could live and work underwater.

The team had no trouble physically adapting to their undersea hangout; in fact, they spent several hours a day diving in the surrounding sea up to 85 feet deep. Thus, they proved that once people were submerged beyond a certain depth, and their bodies saturated with the same nitrogen gas that could cause the bends on a rapid ascent, it didn't matter how long they worked or remained there because the decompression time resurfacing was still the same.

However, the mental stress of underwater living was another matter. Falco was particularly edgy because outside divers and phone calls were disrupting his concentration. He only relaxed once Cousteau and his surface team began limiting contact with Conshelf I.

Nevertheless, the success of this first undersea habitat convinced Cousteau that people could indeed live underwater. He then embarked on an even bolder plan. Conshelf II was a much bigger habitat designed to support five "oceanauts" for a month in four buildings and assorted smaller structures. One of these was a deep diving cabin where two divers could stay for a week.

The main unit in Conshelf II, anchored at 33 feet deep, was Starfish House. This air-conditioned living quarters included a kitchen, dining room, laboratory, and a darkroom where photos could be developed. Other buildings included storage shelters for the diving saucers and anti-shark cages. Conshelf II would be placed on a ***coral reef*** in the Red Sea.

Despite several problems getting Conshelf II up and running safely, the experiment turned out to be a success. Moreover, Cousteau's documentary film about the undersea habitat, called *World Without Sun*, would win him his second Academy Award.

Conshelf II also sparked a team of American oceanographers and divers to build Sealab I, a U.S. Navy undersea habitat where four divers lived and worked for over a week 193 feet below the surface. However, Cousteau's ultimate dream of eventually colonizing the ocean turned out to be only a dream. Even after he established Conshelf III in 1965—where six divers lived for three weeks 325 feet down—Cousteau was disappointed to learn that his efforts didn't lead to increased funding or demand for his technology.

By the 1960s Jacques Cousteau was the best-known explorer of the ocean, thanks to his award-winning documentary films and his television specials.

Treasure Quest 5

For much of his life, Jacques Cousteau had been searching for ways to pay for future research projects. Even before a current expedition was complete he had to worry about what to do next and who would help pay for it. But in the mid-1960s, thanks to his skill at undersea filming, a few visionary people realized Cousteau's adventures might appeal to the general public, who until then had little exposure to the undersea world.

First, he convinced the National Geographic Society to produce a film about the Conshelf III experiment. With help from a Hollywood film editor, the society showed the hour-long film on national TV in the spring of 1966. The special was called *The World of Jacques Cousteau.*

Encouraged to produce more undersea adventures aimed at American audiences, Cousteau made a deal in 1968 with the ABC television network to produce a dozen 60-minute specials over a three-year span. Little did Cousteau realize that this TV series would make him a household name in America, and introduce millions of television viewers to the wonders of the undersea world.

Of course, the business of making television shows required *Calypso* and her crew to spiff up their image. The ship was retrofitted with the latest gadgetry, like closed-circuit television cameras. Also, the divers were asked to wear sporty black wetsuits with bright yellow stripes to match their helmets and tanks. They also began using underwater scooters mounted with lights and cameras to make filming easier, and they skimmed above the water in small inflatable boats called Zodiacs.

Cousteau's first show in the series was about sharks, animals that had long inspired fear and dread among sailors and ocean bathers. The show was a hit, even though it demonstrated that while sharks are unpredictable, most are not menacing man-eaters.

Another filmed expedition took place in the Caribbean Sea. Although parts of the Caribbean are very deep, it is most well known for its far shallower aqua-blue waters and breathtaking coral formations. Though they're beloved by snorkel divers because they attract myriads of tropical fish, coral reefs have been cursed by sailors through the ages because they have caused so many shipwrecks. Certain types of coral can tear open a ship's hull as easily as a pin pops a balloon.

The reefs of the Caribbean were particularly threatening to ships in the 16th, 17th, and 18th centuries, when mariners did not have the sophisticated navigation equipment that sailors use today. In addition, fierce Caribbean storms called ***hurricanes*** could send even the most seaworthy vessel and crew crashing into a coral reef.

Reefs are created underwater by plant-like animals called coral. When they die, they leave hard, stony skeletons. As the coral continues to grow upon dead ancestors, it eventually forms huge limestone reefs.

One of the most treacherous reefs in the Caribbean is known as the Silver Bank. It is just northeast of the island of Hispaniola. According to Cousteau, this reef was home to more sunken ships per square mile than anywhere else on earth. Many of these lost ships contained riches beyond belief: silver, gold, and precious jewels. The valuables had been plundered by Spanish colonists in New Spain (Mexico) and South America hundreds of years ago. They had been placed on large ships called ***galleons*** for the voyage to Spain. However, many ships filled with treasure never made it out of the Caribbean, especially during the hurricane season.

Cousteau intended to look for the *Nuestra Señora de la Concepción*, a Spanish galleon that had ***foundered*** in a hurricane in September 1641 with 525 passengers and crew. Only 200 people survived. Because of its riches, treasure hunters in the 17th and 18th centuries had salvaged some of the *Nuestra Señora*'s vast wealth–silver and gold ingots from Peru and Mexico, jewels from Colombia, and pearls from Venezuela–using native divers to strip what they could from the wooden hulk. What remained to be seen, however, was whether Cousteau and his crew of would-be treasure hunters could even find and bring up the remaining treasures of the Spanish galleon, which most likely was buried by several centuries' worth of coral.

A diver encounters an eel while exploring a coral-covered galleon off the coast of Cuba. The Caribbean is a fertile spot for divers, for many Spanish treasure ships sank in storms there during the 16th and 17th centuries.

Treasure hunter Remy de Haenen, a passenger on *Calypso*, had researched every old letter or record about the wreck that he could find. He thought he knew where the ship's remains had settled.

However, just reaching the search site turned out to be an arduous task. It was surrounded by acres of dangerous coral forests, some barely below the surface. Cousteau and his men were forced to use explosives to blast off the tops of several coral formations just to get *Calypso* safely near the site.

They began searching the ocean floor for the *Nuestra Señora de la Concepción*'s rich cargo, using a powerful suction pipe similar to a giant vacuum cleaner. They also began breaking loose large pieces of coral–some weighing several thousand pounds–and hoisting them aboard the *Calypso* with winches. The blocks were then broken apart with sledge hammers and examined for ***artifacts***.

Initially, the men were thrilled with the prospect of finding sunken treasure, and they worked tirelessly for hours on end in wilting heat. However, the sea was reluctant to give up its riches. Except for a golden cross and a silver belt buckle, the metal most commonly recovered by Cousteau's crew was iron, in the form of rusting cannons.

Eventually, after a month and a half working on the Silver Bank, the men aboard *Calypso* realized that they hadn't even been diving on the wreck of the *Nuestra Señora de la Concepción.* Instead, after recovering several dated iron weights that had accompanied a ship's scale, they discovered that the relics they had been retrieving came from a ship that sank nearly a century after the Spanish galleon went down. Cousteau had little choice but to wind down his undersea treasure hunt and reflect on what his team had gained.

Certainly, it wasn't enormous wealth. Some of the objects raised from the seabed went to museums, some to treasure hunter Remy de Haenen, and the rest–mostly broken pieces of pottery, glass, and iron odds and ends–were divided up among the crew. The expedition on the Silver Bank had cost far more in materials and man hours than it ever reaped in profits. However, the men of *Calypso* had completed a breathtaking, colorful film of their efforts, which probably showed viewers exactly what most treasure hunters can come to expect on undersea digs.

In Cousteau's estimation, something else important had been gained: "Certainly, our investment has already been repaid a thousandfold, in happiness, in satisfaction, in the joy of discovery, in the knowledge that we have been able to contribute something to man's understanding of his world and that we have succeeded in opening, however slightly, a door through which other men will follow."

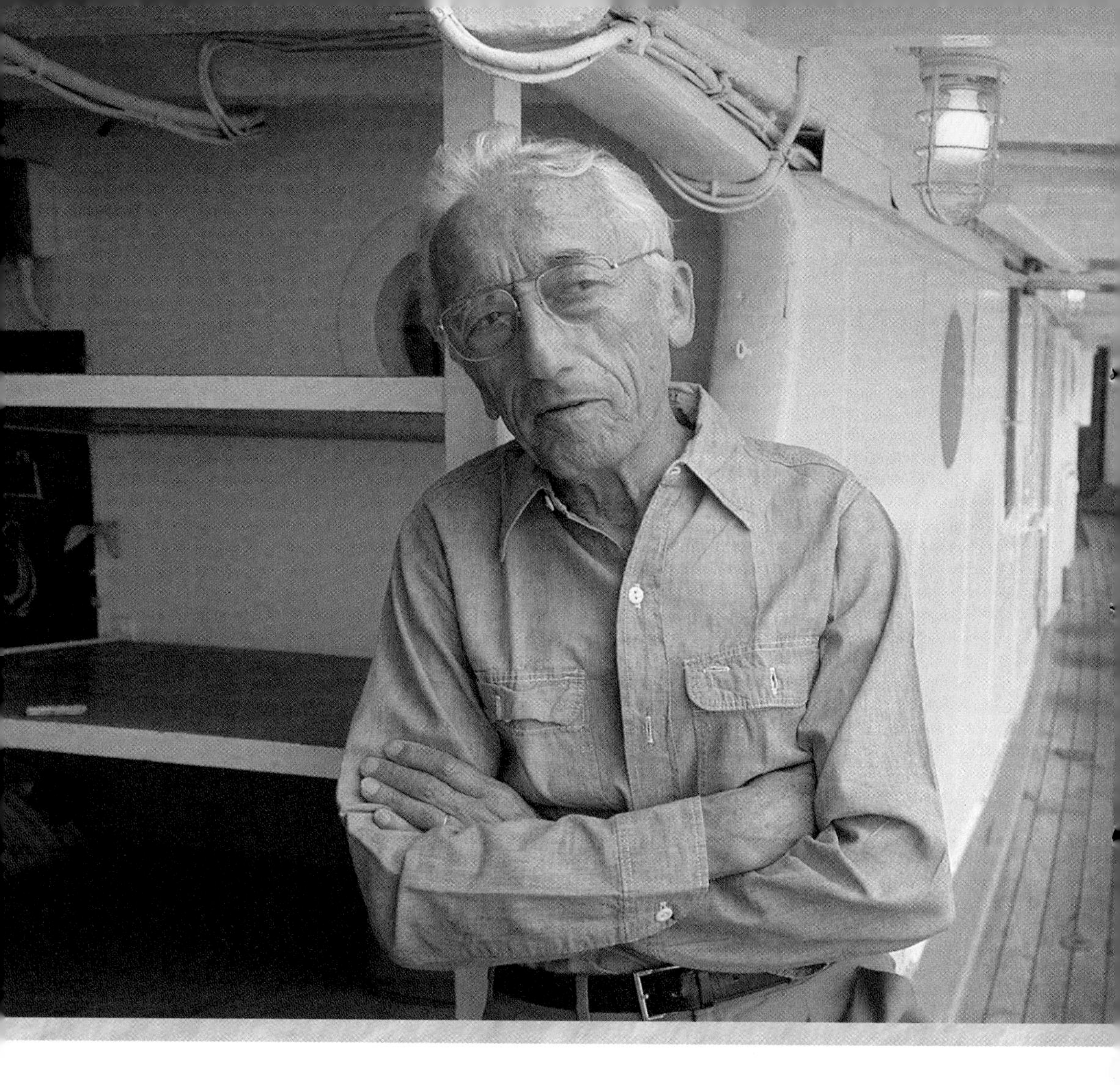

New Horizons

Jacques-Yves Cousteau on the deck of Calypso. *As he grew older, the famous undersea explorer became more involved in environmental causes.*

6

Americans were fascinated with *The Undersea World of Jacques Cousteau.* For eight years people tuned in to watch *Calypso*'s crew dive for sunken treasure, swim alongside enormous gray whales, or bob in kelp beds with playful sea otters. But by the mid-1970s, interest had begun to drop. In 1976, ABC told Cousteau that audiences were no longer tuning in to the four shows a year. His contract was not going to be renewed.

It was the end of a beautiful relationship, but Cousteau was by now an old hand at getting new financial sponsors. He decided to try making films for public television, and promptly lined up Atlantic-Richfield, a large California-based oil company, to sponsor a series, *The Cousteau Odyssey.* The new program did not focus solely on undersea adventures, however. *The Cousteau Odyssey* set out to publicize global threats to humans and their environment: everything from oil spills and overfishing to battles over land management and vanishing civilizations.

At times, this type of work was a family affair.

Philippe Cousteau at the wheel of a sailing ship, with his father standing on the deck behind him. Jacques worked with both of his sons on the long-running TV program The Undersea World of Jacques Cousteau.

Philippe and Jean-Michel Cousteau worked alongside their famous father in various roles. They were fellow divers and filmmakers. Philippe was a pilot, and he even outfitted *Calypso* with a ***seaplane***, hot-air balloon, and hang-glider. This would help the Cousteaus give viewers a bird's-eye view during the *Odyssey* series.

But although Cousteau's two sons shared his love of the watery world, he didn't always get along well with them. Jean-Michel, who had mostly worked behind the scenes during the ABC television series, drifted away from his father for awhile because he felt he was being taken for granted. And for several years Cousteau and his wife, Simone, had little contact with their son Philippe after he married an American model whom they didn't like.

Thus, it was all the more tragic for the elder Cousteau when Philippe died in 1979 during a test flight of their seaplane, *The Flying Calypso.* The plane had just undergone repairs. Philippe was landing the plane on the Tagus River in Portugal when it suddenly flipped over and broke apart, killing him. His grief-stricken father couldn't even bring himself to work after the accident. However, after Jean-Michel agreed to come back and work with his father by directing the Cousteau Society (the non-profit organization that had been formed by Jacques five years earlier), the elder Cousteau resumed his marine exploration projects.

In some respects he had little choice. His public television series had ended, so once again he needed to find funding for his work. This time Cousteau struck an agreement with cable television mogul Ted Turner. Turner, who owned the TBS cable network, shared Cousteau's passion for environmental issues. He agreed to finance a Calypso expedition of the Amazon River and broadcast it on his network.

The extensive two-year expedition began in 1982. It included over 50 Cousteau divers, scientists, and filmmakers. Just the equipment required to make the documentary film was incredible. In addition to *Calypso*, there were floating vessels of all sorts–kayaks, rafts, Zodiacs; assorted jeeps; two amphibious trucks; and a helicopter. The hardware was essential because separate crews were exploring and filming at any given time in Peru, Brazil, Colombia, and Bolivia.

In all, Cousteau's team filmed and produced four shows highlighting the incredibly diverse animals, reptiles, fish, and people of the Amazon River basin.

The programs gave television viewers a rare glimpse into an area threatened by the destruction of its rain forests for farming and development.

The Amazon River is 4,000 miles long, the widest river in the world, and lifeblood to the largest tropical rain forests on the planet. Moreover, it is home to over 2,000 different fishes, a variety unmatched even in the Atlantic Ocean. It is also home to the world's largest snake (the anaconda) and otter, freshwater river dolphins, and the last refuge of several unique native tribes.

In his later years, Jacques Cousteau had become a well-known environmentalist. He often sounded alarms over what he saw as exploitation of the earth's water and land. And in a related effort to boost fuel conservation, Cousteau designed a seagoing vessel named *Alcyone.* The ship featured Turbosails, two smokestack-like structures that took advantage of wind power to reduce diesel fuel consumption by about 40 percent.

Cousteau's pioneering work in the field of marine exploration and environmentalism won him many fans–300,000 of whom are now members of the Cousteau Society–and several notable awards.

Jacques Cousteau's experimental ship Alcyone *sails into San Francisco Bay. The ship was powered by a combination of wind and internal diesel engines.*

Jacques received the Presidential Medal of Freedom from Ronald Reagan in 1985. Four years later, he was inducted into the Academie Francaise, France's highest honor, in recognition of his lifelong contribution to the country's culture.

Even though he was more than 80 years old, the undersea explorer continued to dive, although he mostly stayed in warm waters. Following the death of his beloved Simone in 1990, he remarried and lived in Paris, where he continued to plan future projects. One of them was raising money to design and build a replacement for *Calypso*, which had logged over a million miles and served research purposes well for decades but was now showing its years. (In fact, *Calypso* sank in Singapore Harbor in 1996 after being struck by a barge while awaiting an expedition up the Yellow River of China. It was raised, however, and now is a featured attraction at a maritime museum in La Rochelle, France.)

The new vessel Cousteau envisioned would be outfitted with state-of-the-art navigation equipment as well as cabin space for passengers. In fact, his goal was to seek small contributions from people worldwide to pay for the multimillion-dollar research ship–in effect making the world a shareholder in Cousteau Society expeditions.

The years eventually caught up with the famous marine explorer and ***oceanographer***. On June 25, 1997, he died.

But Cousteau's legacy lives on. Today, the society named after him continues to raise funds to build *Calypso II*. The Cousteau Society says that the environmentally friendly ship, which will cost $40

Divers in a Zephyr examine the sunken Calypso. *The ship went under in Singapore after it was hit by a barge in 1996. Jacques-Yves Cousteau's famed ship was eventually raised, and now is an exhibit in a museum.*

million, will be an "environmental watchdog for the Earth," keeping "her finger on the pulse of the planet with the capability to gather and transmit, from anywhere around the globe, essential information on the condition of our world."

Captain Cousteau, who made it his mission to explore, document, and ultimately safeguard life on our planet, would undoubtedly approve.

Chronology

1910 Jacques-Yves Cousteau is born in St. Andre de Cubzac, France, on June 11.

1930 Enters French Naval Academy.

1937 Marries Simone Melchoir.

1943 With French engineer Emile Gagnan, develops the first regulated compressed-air breathing device.

1948 Salvages the wreck of the argosy of Mahdia with Undersea Research Group.

1950 Purchases *Calypso*, a former minesweeper and ferry vessel, and converts it into a research ship.

1953 Publishes *The Silent World*, his first book.

1954 Convinces France to sponsor *Calypso*'s expeditions.

1957 Documentary film *The Silent World* receives an Academy Award in Hollywood; accepts position as head of Monaco's Oceanographic Institute.

1962 Conducts Conshelf I experiment off Marseilles, France.

1963 Launches Conshelf II test in the Red Sea.

1965 Completes Conshelf III research near Nice.

1968 *The Undersea World of Jacques Cousteau* begins eight-year run on ABC-TV.

1974 Founds the Cousteau Society.

1979 Son Philippe Cousteau dies in plane crash in Portugal.

1982 Helps develop the Turbosail wind-propulsion system.

1985 Wins U.S. Presidential Medal of Freedom.

1989 Inducted into Academie Francaise.

1997 Dies on June 27 at age 87.

Glossary

aqualung–the name for an underwater breathing apparatus developed by Jacques Cousteau and Emile Gagnan. This equipment is also known as scuba, which stands for Self-Contained Underwater Breathing Apparatus.

argosy–a large merchant ship.

artifacts–old man-made objects.

bathyscaphe–a craft built for deep-sea exploration that can be navigated and has a round watertight cabin attached to its underside.

bends–a potentially fatal physical disorder, caused by the release of nitrogen gas bubbles in the body after a rapid decompression.

coral reef–a large, hard underwater ridge made up of the stony skeletons of sea creatures called coral.

decompress–to undergo release from water pressure.

documentary film–a movie that takes a factual and objective look at its subject.

finance–to provide money needed to accomplish a goal.

founder–to sink below the surface of the water.

galleon–a heavy, square-rigged ship used by the Spanish for war and commerce from the 15th to the 18th centuries.

habitat–a controlled physical environment in which people can live under conditions that would otherwise be impossible (for example, under the sea).

hurricane–a tropical storm with high winds.

mine–an explosive that is placed in water or in the ground, and set to explode if touched or disturbed.

minesweeper–a warship designed to remove or neutralize mines.

navigation–the science of directing the course of a seagoing vessel, and of determining its position.

nitrogen narcosis–a state of euphoria that occurs when nitrogen in normal air enters the bloodstream at approximately seven times the normal atmospheric pressure. This condition is also called "rapture of the deep."

nonprofit organization–a group dedicated to scientific or humanitarian causes, rather than making money.

oceanography–the science of studying the ocean and the creatures that live in it.

principality–sovereign territory that is ruled by a prince.

salvage operation–an attempt to remove valuable items or goods from a wrecked ship.

seaplane–an airplane designed to take off from and land on water.

water pressure–the force, or pressure, that water exerts on items it surrounds. Water pressure increases with depth.

Further Reading

Cousteau, Jacques-Yves. *The Silent World.* New York: Harper and Brothers, 1950.

———. *Jacques Cousteau: The Ocean World.* New York: Harry N. Abrams, 1979.

Cousteau, Jacques-Yves, and Philippe Diole. *Diving for Sunken Treasure.* New York: Doubleday and Company, 1971.

The Cousteau Society. *An Adventure in the Amazon.* New York: Simon and Schuster, 1992.

Dutemple, Lesley A. *Jacques Cousteau.* Minneapolis, Minn.: Lerner Publications, 2000.

Hopping, Lorraine Jean. *Jacques Cousteau: Saving Our Seas.* New York: McGraw-Hill, 2000.

Konstam, Angus. *Historical Atlas of Exploration.* New York: Facts on File, 2000.

Markham, Lois. *Jacques-Yves Cousteau: Exploring the Wonders of the Deep.* Austin, Texas: Raintree Steck-Vaughn, 1997.

Index

Picture Credits

page

3:	Reuters/Archive Photos	32:	Everett Collection
6-7:	AP/World Wide Photos	34:	AP/Wide World Photos
9:	Jonathan Blair/Corbis	39:	AP/Wide World Photos
12-13:	Everett Collection	42:	Everett Collection
15:	AP/Wide World Photos	47:	Arne Hodalic/Corbis
18:	Lambert Studios/Archive Photos	50-51:	Bettmann/Corbis
22-23:	Everett Collection	52:	Everett Collection
29:	AP/Wide World Photos	56:	AP/Wide World Photos
31:	Stephen Frink/Corbis	58:	AP/Wide World Photos

Cover photos: (top) Arne Hodalic/Corbis; (right) Everett Collection

ROGER KING, a former newspaper columnist, has written several books for young adults. He lives with his family near Pittsburgh, Pennsylvania.